MW00890407

From One To Infinity With Synergy

by

Wallace A. Johnson

From One To Infinity With Synergy

From One To Infinity With Synergy

Published by Wallace A. Johnson

Printed and Distributed by CreateSpace, a division of Amazon.

www.spaceshipdewaj.com

From One To Infinity With Synergy

Dedicated to my wife Doris,

for her undying synergy.

From One To Infinity With Synergy

APOLLO PROJECT).

2. THE TRAGIC FIRE OF APOLLO 1

CURRENT EVENTS

1. SENIOR CITIZENS, COMPUTERS AND THE INTERNET

2. A MODERN DAY TRAGEDY

From One To Infinity With Synergy

FOREWORD

My logo is: *12 Infinity With Synergy*. It stands for *One To Infinity With Synergy*. Now you understand. But why did I choose that? Let me explain. My blog tries to give you a glimpse about the journey I have been on since the day I was born on April 18, 1925 in the small town of Taft, California. I am 93 now in the year 2018, and to be frank I never expected to live to the year 2000. But it was a goal of sorts because I wanted to live to the ripe old age of 75 and bring the millennium in. To my surprise I not only made it but I'm still going strong! It is said, you come into this world alone, and you leave it alone. I don't know whether that is true or not, but I do know this. I have accomplished and done many things these 93 years, but I did NOT do it alone. My life has been blessed because from my early recollection, other human beings influenced me or changed my environment in some way that ultimately but surely molded my thoughts and actions such that the person I am is the result. I will attempt to describe this journey and point out those with whom I have interfaced with and how

From One To Infinity With Synergy

they influenced my actions at the time. So I see my life as the culmination of a team effort using a Synergistic approach. Many contributed and the end result is a sum which is greater than the summation of the individual parts. I am who I am because so many played a part in this Saga of which I am proud.

I am tying my current effort of chronicling this journey by cohesively putting together my thoughts ending with *I Have Lift Off* and *Spaceship DEWAJ* **(a Daring Enterprise With A Journey)**. Be my Co-Pilot, join me on my journey from One To Infinity With Synergy.

From One To Infinity With Synergy

RESOURCE

Wallace A. Johnson MBA, Commander of Spaceship DEWAJ, was a Test Pilot on the Apollo Project working for North American Aviation in Downey, Ca. during the 1960's. Now at the age of 93 this April 2018, he has his own spaceship and is offering many programs of interest to all. Many items are available for the taking. He is encouraging Guru's to participate with Joint Venture possibilities. The main thrust is to enlist the aid of others in a manner which will ensure that Spaceship DEWAJ remains in orbit long into the future. Currently Free Lifetime Charter Memberships are offered. At the age of 93, the Senior Navigator Test Pilot's time is short and his Spaceship DEWAJ will be his legacy to his family, friends, and the Internet. The Commander requests you consider joining him as an

From One To Infinity With Synergy

Associate Co-Pilot on his Spaceship DEWAJ, a Daring
Enterprise With A Journey. Welcome aboard.

From One To Infinity With Synergy

The Early Years

THE BIRTH OF SPACESHIP DEWAJ: A DARING ENTERPRISE WITH A JOURNEY

As a former Test Pilot on the Apollo Project, I have always dreamed what it would be like if I could have had my own Spaceship. Through the medium of the Internet, I can now fantasize. This Blog will be part of that fantasy and will tell you how Spaceship DEWAJ came to be.

SOME PERSONAL HISTORY

At the age of 93, time flies and the days shorten. Widowed, and without children, I can see that those who through the years have asked me to put into words, my rambling thoughts, inspirations, ideas, mistakes, adventures, travels, etc. have a point. Not that anything I might say will change much, for I don't think that anything I might say will have that much influence, but in fact it has been in the back of my mind for some time now. So here it is at last.

From One To Infinity With Synergy

I'm one of those who live in California that was actually born here. My birthplace was Taft, Ca. and my birthday is 18 April 1925. Taft, California is oil refinery country. My father worked for Standard Oil Co. and before I was six months old, my father returned to Havana, Cuba, where he had formerly married my mother. As the Superintendent of the Beloit, refinery in Havana, we lived on Standard Oil Co. refinery grounds. I had a brother Willis, and a sister Wanda, who were born there. We lived well, and my childhood memories are those of a loving father and mother who doted over their children, especially me. However, those of you that remember your history will recall a revolution took place in Cuba in 1932. With Cuban soldiers bivouacked on our front lawn to protect the refinery, it didn't take my father long to decide that the safety of his family was paramount. So in late 1932 we returned to Houston, Texas.

Talk about jumping from the frying pan into the fire! The depression years were upon us, and times were rough. This was especially the case, when due to an accident at the refinery my father had an untimely death. That left my mother alone, with no skills other than being a good wife and mother with three young and quite often hungry children. As I said, times were rough.

Many people throughout my life have had an impact on me culminating in the person I am today. The input of these individuals worked in a synergistic way and the result is a number greater than the sum of the individual parts. I am therefore a greater man because of the help and influence of others, to all of them I am grateful.

I Flight Instructed with the business name of DEWAJ Flying Service. DEWAJ stood for "Doris Elizabeth & Wallace A Johnson". When I decided to start my journal, I called it my Spaceship DEWAJ with the acronym now standing for a "Daring Enterprise With A Journey." So far, the journey of my life has gone on for over 93 years, and although I am in remission fighting the "Big C". (Stage IV Melanoma) and Prostate Cancer, my hopes are high, and with all good luck I intend to hang around a few more years. You are invited to join me on my journey.

From One To Infinity With Synergy

My logo says 12 Infinity With Synergy. Let's go from One to Infinity With Synergy on Spaceship DEWAJ. Welcome Aboard!

THE NEW DEAL & THE ALPHABET

It would help you immensely, if you were my age or thereabouts, to understand what I am about to tell you. I remember distinctly hearing the inaugural speech given by the new elected president Franklin Delano Roosevelt. It was notably remembered by historians as the message where he said, "We have nothing to fear but fear itself." I could be wrong about the exact words, but that was its meaning. That famous quote was followed by his saying that the nation needed a **New Deal**. and that leads me to the Alphabet.

Government agencies are notorious for acronyms, and I can understand the dilemma a new agency has when it has to tie its name with its function. I can relate to the problem in my own personal way. Let me digress for a moment. Many of

From One To Infinity With Synergy

you know that my wife's name was Doris Elizabeth. So I tied her first two initials DE, into my initials WAJ, and came up with DEWAJ as an acronym for my DBA (Doing Business As) license. Also, DE in Spanish stands for *of* and it tied in well with the WAJ portion. So for years I used DEWAJ in many ways, including the registration of a Star in the constellation of Sagitta near Sagittarius to be designated as DEWAJ. I did that as a present to my wife on our wedding anniversary. When I decided to get my own web site, I wanted to use DEWAJ, which explains my doing business as DEWAJ Synergy International. That has worked well for some years now and I continue to use it to this day. However, when my wife died, I decided to really start concentrating on my web site in earnest. So I decided to have my own Spaceship and naturally I wanted to call it Spaceship DEWAJ. So far so good, I had used DEWAJ OK, but to tie the acronym to it was a problem. It took some time, but all of a sudden it hit me like a bolt of lightning. DEWAJ, a *Daring Enterprise With A Journey*. and it fit perfectly. The idea was to have a Spaceship which would use Synergy to accomplish its mission to remain in orbit long after I am gone. By including others, we would

From One To Infinity With Synergy

use synergy and end up with an end result whose sum would be greater than the sum of the individual parts. I thought it to be a brilliant idea as an acronym and I'm real proud of it. So now my job is to make the acronym well known on the internet. That's a daunting task, and I know I can't do it alone. But if I can get any exposure at all, and with help from others who will be either Passive Passengers or Active Joint Venture Partners, I know that using Synergy, it can be done. Enough of that, to get back to *The New Deal & The Alphabet*.

New administrations whether government or private, are prone to come in with a new broom and give the place a clean sweep as it were. Roosevelt did this as well. So to give us all a **New Deal** required the building up of vast bureaucracies, all with high sounding names. Bureaucratic Alphabetization which was nothing new in government circles, really took off. If you can remember some of these, it will date you for sure, but I don't care, in my case, you know I'm 80 right now, and I remember it with some nostalgia. Just to name a few. Do you remember the N.R.A. (National Recovery Act), how about the W.P.A. (Work Projects

Administration), you can see the results of many projects completed under the W.P.A. to this day in the form of libraries, bridges, dams etc. The C.C.C. was another favorite one, especially for young kids who couldn't find employment anywhere else. Here is a list I found and what they represented including some important dates.

United States bank holiday, 1933: closed all banks until they became certified by Federal reviewers.

Abandonment of gold standard, 1933: allowed more Money to be put in circulation to create a mild inflation.

Civilian Conservation Corps (CCC), 1933: employed young adults to perform unskilled work for the federal government.

Tennessee Valley Authority (TVA), 1933: a government program that ran a series of dams built on the Tennessee River.

Federal Emergency Relief Administration

(FERA), 1933: provided breadlines and other aid to the unemployed.

Agricultural Adjustment Act (AAA), 1933: paid farmers to not grow crops.

National Recovery Act (NRA), 1933: created fair standards in favor of labor unions.

Civil Works Administration (CWA), 1933: provided temporary jobs to millions of unemployed.

Public Works Administration (PWA), 1933: employed middle-aged skilled workers to work on public projects, cost $4 billion.

Federal Deposit Insurance Corporation (FDIC) / Glass-Steagall Act: insures deposits in banks in order to restore public confidence in banks.

Securities Act of 1933, created the **Securities and Exchange Commission (SEC)**, 1933: codified standards for sale and purchase of stock, required risk of investments to be accurately disclosed

Indian Reorganization Act, 1934.

Social Security Act (SSA), 1935: provided financial assistance to: elderly, handicapped, delinquent, unemployed; paid for by employee and employer payroll contributions.

Works Progress Administration (WPA), 1935: a reiteration of the PWA, created useful work for skilled workers.

National Labor Relations Act (NLRA) / Wagner Act, 1935: granted right of labor unions to exist.

Fair Labor Standards Act (FLSA), 1938: established a maximum normal work week of 40 hours, and a minimum pay of 40 cents/hour.

Although Roosevelt was attacked then as now for that matter, you would be hard pressed to get rid of some of those programs that are still with us today, even though they may have a different name.

From One To Infinity With Synergy

Social Security is one example. However, I want to point out some incongruities that to this day I don't understand. Take for example the A.A.A. where farmers were paid NOT to grow crops. I understand the law of supply and demand and the need for a market to exist, but I can remember kerosene being poured on piles and piles of oranges then having them set on fire to destroy them, all the while, I would give my right arm just for the chance to savor the beautiful fruit denied me because I couldn't afford to buy it. I can also remember agents of the government rounding up many head of cattle into a lot and indiscriminately shooting them dead. You would think, hoof and mouth decease, but it wasn't that at all, just that there were TOO MANY cattle on the market and to sustain a price on cattle, you had to have a scarcity. All the while there were an awful lot of people who weren't eating much meat believe me, and I was one of them.

These are just a few of the things that I remember occurring during that time which made me very cognizant of social conditions and some of the inequities that existed, many for no reason at all that made any sense to me. Somehow, my family

From One To Infinity With Synergy

survived. In my next entry to my blog, I'll let you know how I did my best to understand and survive as well.

From One To Infinity With Synergy

Starvation None, Malnutrition Plenty!

As I was growing up during the 30's, I rapidly matured, not only physically but mentally. I acquired a "Social Conscience" early on. I mentioned in my last entry, how bewildered I was to see good food being destroyed merely to sustain a price. It didn't make sense to me then and it doesn't make any more sense now. However social planners always have altruistic end results in sight and there is no doubt that it could all be explained, at least in their minds, as having an end result which was better than not interfering at all. I can remember that the unemployment figures were 25 million or more. Able bodied men begging for work, and those that didn't find it were left with begging period! Something happens to a person's dignity and self-worth, when

From One To Infinity With Synergy

he is forced by circumstances beyond him, to literally have to beg to insure his survival. Worrying about oneself is one thing but to realize that one is responsible for others survival as well, is incomprehensible unless you have experienced it yourself. It's an animal of a totally different color. A day's pay at back breaking labor was $1.00 and that's all one could sell his labor for, and some were willing to do it for less. You heard it said then and it's still repeated today, "Nobody died from starvation in the 30's" and perhaps there is some truth to that, but there sure was a lot of malnutrition and the thought comes to mind that maybe malnutrition is just another nice way of saying slow starvation.

I can remember clearly that I was given a physical at the age of 11 and the doctor conducting my examination was an elderly lady perhaps in her late 50's. I have no idea what the real circumstances were but I remember that it had to do with the fact that my mother was applying for some assistance in getting milk for the family. With three children, 11, 9, and 7 there just was no way that she could afford to buy milk for all three. So we were doing without milk, and it was showing in our physical condition.

From One To Infinity With Synergy

The doctor was quite disturbed and was arguing with some functionary about the fact that we had not been included in a program the City of Houston had where milk was being made available for those who qualified. The milk was being distributed at local fire stations throughout the city, but my family was not on the list. The doctor in a large and commanding voice, was heard by me to say "I don't give a damn what you say, I said these kids will be given a quart of milk a day period, now do it" and the social worker made out the necessary forms so that our family would be eligible for the free milk. I have no ideas who that doctor was, but I know that there are angels that walk this earth, and in reflecting on that moment now, I know she was one of them. Can you imagine an angel cursing? Well under certain circumstances they do, and I'll swear to it. It's hard to imagine an angel that is pissed off; believe me, when they get that way, you pay real close attention. So we got our milk.

I used to take two buckets with me and I would go to the local fire station which was quite some distance from where we lived and I would load those buckets up with jars of milk. It would be enough to take care of our needs for a whole week, whereupon I would return and repeat the process. Milk in those days was not pasteurized and was called "Raw" milk. The neck of the bottle was built different than the bottles we had later on when it was homogenized. It had a bulge at the top of the bottle, and you could see the line where the cream had risen to the top and was laying in that section of the bottle where the bulge was. The majority of the time we got skimmed milk which had its cream removed for making butter, but it still had a lot of nutritional value skimmed or not. Another thing I remember was that the cap on the bottle was a round cardboard lid with advertising on it. Today, the whole container is made of a cardboard composite. Do you remember when the milkman delivered milk in a glass bottle? I don't think I have ever tasted milk since that tasted like it did then. Nostalgia, nostalgia, how beautiful it is.

One summer, I think it was 1934 or 35, I can't remember, it was decided that I should go to spend some time with some friends who had a farm in the Navasota, Hempstead area. The thinking was that I would get three square meals a day and perhaps put some weight on. I was not starving, but I was experiencing some malnutrition, so off I went. I

From One To Infinity With Synergy

remember, Aunt Fanny, (No relation), cooking pancakes and feeding the work hands sitting at a long table. The table could handle twenty people at one time, it was so immense, and she would cook pancakes by the dozen. They made sorghum molasses on the farm, and I can still see them slowly moving the mixture as it cooked on a roaring fire ever mindful of the fact that at some point it would all of a sudden get to the right consistency in thickness requiring it being removed from the fire immediately. Going past a certain point could also cause it to burn, and I have had the pleasure of having to put burnt molasses on my pancakes more than once. It couldn't be thrown away just because it was overcooked and burnt. And woe to the person whose job was to keep it moving in the ladle, who let it get past that certain point. I knew my day for stirring was coming, and sure enough, I sweated that ordeal with the certain knowledge that I wasn't going to mess up on my assigned task. When the liquid is first put into the vat, it's not unlike water in its viscosity, but as it cooks it slowly thickens and when I noticed the resistance to the pole I was stirring it with, I yelled like a banshee, because I knew too much longer and it would be burned. When you realize that if you burn the molasses, you are in for a real lambasting and your life threatened, you keep a close eye on what you are doing. I was lucky, I stirred the molasses many times, but I never let it burn. My guardian angel has been with me from an early age, even to this day.

Unless you're my age or thereabouts, you don't have any idea what I'm talking about when I say "Picking Cotton." In the first place, it's most likely 110 degrees in the shade. You are dragging a canvas bag that is about 6 to 8 feet long and a diameter of a large wash bucket. The bag has a large loop at the opening which you put over your shoulder, and you literally drag that bag as you reach over and pick the boll of cotton and drop it into the bag. The protective leaves that surround the boll of cotton has sharp prickly points on it, and in no time at all, your fingers are bleeding from the pricks you endure. Not much fun I assure you. A grown man was expected to fill that bag up and with a lot of cramming could get many pounds into the bag. I don't remember exactly but it seems to be that all men strived to get 100 pounds for the day into the bag. And you would be immediately fired if they caught you, but more than one man who had to relieve himself would urinate into the bag. Anything to make that cotton heavier! Can you imagine how many bolls of cotton a man has to pick just to get one pound of cotton? I picked cotton right along with the rest that summer. I was shooting for ten pounds, and I made it. The bags were weighed on a devise where weights were suspended on a fulcrum to determine the weight of the cotton. All crude by modern standards, where a human hand no longer touches the boll of cotton, but rather a mechanized cotton picker comes along and strips the cotton from the plant, doing it thousands of times as fast and cheaper in the long run. Talk about Technological Unemployment and the machine replacing manual labor!

But that's another story.

MY SPACESHIP DEWAJ: A Daring Enterprise With A Journey

SOME PERSONAL HISTORY

At the age of 93, time flies and the days shorten. Widowed, and without children, I can see that those who through the years have asked me to put into words, my rambling thoughts, inspirations, ideas, mistakes, adventures, travels, etc. have a point. Not that anything I might say will change much, for I don't think that anything I might say will have that much influence, but in fact it has been in the back of my mind for some time now. So here it is at last.

I'm one of those who live in California that was actually born here. My birthplace was Taft, Ca. and my birthday is 18 April 1925. Taft, California is oil refinery country. My father worked for Standard

From One To Infinity With Synergy

Oil Co. and before I was six months old, my father returned to Havana, Cuba, where he had formerly married my mother. As the Superintendent of the Beloit, refinery in Havana, we lived on Standard Oil Company refinery grounds. I had a brother Willis, and a sister Wanda, who were born there. We lived well, and my childhood memories are those of a loving father and mother who doted over their children, especially me. However, those of you that remember your history will recall a revolution took place in Cuba in 1932. With Cuban soldiers bivouacked on our front lawn to protect the refinery, it didn't take my father long to decide that the safety of his family was paramount. So in late 1932 we returned to Houston, Texas.

Talk about jumping from the frying pan into the fire! The depression years were upon us, and times were rough. This was especially the case, when due to an accident at the refinery my father had an untimely death. That left my mother alone, with no skills other than being a good mother with three young and quite often hungry children. As I said, times were rough.

From One To Infinity With Synergy

LIFE IN THE 30's (Continued)

It was 1935 when my father died. I was 11. I had a Sister (Wanda) who was 9, and a brother (Willis), who was 7 and the youngest of the three. Mother did her best to hold the family together and I did my best to take my father's place since I was the oldest of the three children. Mother ended up working as a seamstress for the WPA (Works Progress Administration). Franklin Delano Roosevelt was the president of the United States and his "New Deal" with its alphabetism of Federal Agencies, was an attempt at government to solve the ensuing problems associated with the results of the market crash of 29. Were it not for some of those programs, I really don't know what our family would have done. My mother had two brothers and a sister that lived in Houston and I remember that my uncle Raphael moved in to help mother with the rent. It was the sum total of $12.00 a month for a two bedroom apartment. I still remember the address as 1213 Decatur St. and I was attending Dow Junior High School.

But let me go back a little and fill you in on some of the little things that I remember during that time which made such an impression on me as a young boy. Before my father died, he arrived in Houston with some savings, and I remember him going up to Navasota and the Hemstead area of Texas to look at some land to buy thinking that maybe if things really got bad he could somehow live off the land if he had to. I can see him now reaching down and picking up some of the loam and crushing it in his hand and actually smelling it, as if to get a feel for its quality. I know now that he was trying to determine whether the land would be fruitful and able to sustain a family of five. Thank God he decided that there was no way the land could support us. It too was worn out and useless so we returned to Houston where he ultimately bought the Texaco filling station.

While in the Navasota area, we stayed with my Uncle Fernando and his wife Malcom whom he had recently married. Her father had a small farm, and it somehow fed us all and kept us together. Here is where I got an appreciation for the land and the other animals we shared our meager existence with. You have to see how the lower animals struggle to survive the visitations we as humans bring to them. You have to see animals give birth, then those born become beast of burden followed by their dying, to marvel at the wonder of it all. And you don't have to be too sharp of mind to realize that somehow we as animals fall into the same circumstance as we live our human experience. To survive, we as humans have somewhat of an advantage due to our higher intelligence, but sometimes I am saddened when I realize how little we think of the lower animals and the way we treat them. I remember vividly one occasion when a poor mule was being used to drag a scupper which was used to take out the muck and mud at the bottom of what was actually a drinking pond for the animals. It was low on water and the mule would walk through the mud and the scupper would take out the mud. Unfortunately, the poor mule got bogged down in the mud up to his belly, it couldn't move out on its own much less pull the scupper through the muck and mud. The farm owner, who at the time was drunk and mad with rage at the mule, took a chain and started to beat the poor animal. The animal was crying out in terror at the pain inflicted on him and I in turn who was watching this bestiality was crying out loud as well. I couldn't believe what I was seeing. Undoubtedly the drunkard thought that if he inflicted enough pain on the animal, the animal would somehow miraculously gain sufficient strength to extricate himself from its predicament. It ultimately required chains being wrapped around the neck of the animal and a string of men literally had to drag the poor beast out. How that animal survived his treatment at the hands of that man is beyond me, but I think of that episode often when I see someone mistreat a lower animal for no reason at all. I believe it is a sad reflection on the human conditions when we treat animals the way we do only because they are "lower" animals. If we allow our children to mistreat animals, then why should we be surprised when those children become adults, and have little regard for the feelings of humans? They have become desensitized to those feelings necessary to understand the sanctity of a living being, whether a lower animal or a higher human animal. There are too many of us that have forgotten that humans are animals too. It's no wonder it is so easy to have men kill each other in this thing called "war." The lower animals kill as a necessity for survival, whereas man kills in many cases for no reason that has anything to do with survival. As I said, it's a sad reflection on the human condition.

As I said previously, when by father returned to Houston he bought a Texaco filling station as an investment. My Aunt Mary's husband Joe Ferguson ran the station for him. It was on Leeland Avenue, and I remember it well. However, as you can imagine, any investment during those years quite often went sour. Nothing seemed to work out and in short order the small stake he brought back to the States after leaving Cuba, dwindled and he found himself just as broke as many others were in those dark days. It wasn't too long after, that he had an accident at the refinery. His health suffered and his spirit as well. His body slowly gave in and he died a rather sudden death. His death certificate says he died of Uremic Poisoning due to Kidney failure. In those days, a lot of people died for reasons unknown. Pneumonia was often the culprit, and in my father's case I don't really know what the cause of death was exactly. I believe, it was more from a broken heart than anything else. We believe he was 49 at the time.

Things were rough in the 30's. My Mother did all she could to keep the family together after my Father passed away. I was 11 and doing rather well in school, but it wasn't without the help of some people I will never forget. In particular, my music teacher Ms. Norris. I was laboring to master the violin and cello, and as it happens from time to time, one of my strings would break, and there I would be. I couldn't afford to buy strings, but somehow, the next time it came to be at practice, Ms. Norris would hand me the necessary string. I know she couldn't afford those extra expenses herself, but as a substitute teacher currently, I know that teachers quite often supply their students with necessary items which students can't afford themselves. Teachers are a special lot, at least the good ones are, and Ms. Norris was one of those. She was very instrumental in my appreciation of music in all of its forms and she was a very important part in the eventual formulation of the person I am now and have been in the past. She was a kind soul and I will never forget her. Another person I will never forget

From One To Infinity With Synergy

was a Mr. Ben Duggan. He was an executive for an insurance company there in Houston, Texas. He was notified about my particular circumstances by my home room teacher Ms. Fern Smith. He took it upon himself to more or less adopt the family, me in particular since I was the older of my siblings. I remember on one occasion being taken to a professional baseball game for the first time. It was the Houston Buffaloes, and we had seats right behind home plate. I'm not much on sports, but I was impressed with the whole thing and now realize how kind he was to me and my family. I saw him as a good man who was just trying to do a kind act out of the goodness of his heart. During the war, while I was overseas, my mother wrote me inform me of his passing by sending me a clipping of his obituary which was in the Houston Chronicle. I still have that clipping believe it or not. He was a special person, put on this earth for a special reason, and I now realize I was part of it in a small way. I came to find out that my family wasn't the only one who he helped and he was highly respected in the Houston business community for his various charities. I mentioned my home room teacher Ms. Fern Smith. If there ever was an angel who walked the earth, she was one. I had her in the third grade while I was attending Hawthorne Elementary School. Let me digress for a moment. While living in Cuba, I had attended school on refinery grounds and was in the third grade. All classes were taught in Spanish. I was lucky because my mother would speak to me in Spanish and my Father would speak to me in English. I always responded in Spanish. So I understood both, but my main tongue was Spanish. When I started school in Houston, with my language problems, I was put in the first grade. I picked up my English speaking abilities rapidly and within a year's time found myself in the third grade. So I was at the proper grade level for my age in no time. Ms. Smith was my home room teacher at the time my father passed away and took an immediate interest in me and my family. She became very close to the family and would visit from time to time insuring all was well and that we were being cared for. She was the person who told Mr. Duggan about my family and I kept in constant contact with her with personal visits and correspondence through the years My first letter from her came when she wrote on the passing of my father. I still have that letter. I wrote to her and visited with her on many occasions through the years until her passing when she was in her 90's. She was an old maid, never married, who was very frugal, but somehow amassed an impressive portfolio of stock holdings and real estate, which came to a sizable sum. I remember clearly, during WWII, I was on leave one day, and I visited her. She was having her lawn mowed by a young boy who lived next door to her and she introduced me to him. I remember the occasion well, because although I was 19 or so at the time, he made an impression on me, for it was obvious Ms. Smith thought well of him. Ms. Smith dabbled in oil paintings and was quite prolific with pastoral scenes and flowers. During one of my letter writings to her in the 90's, I received a letter from an attorney by the name of Kennedy. He informed me of her passing, that he was her lawyer and was the executor of her estate. He informed me that in her estate, she had mentioned me by name, and that I was to have the choice of three of her paintings. Not only was I shocked to hear of her passing, but the fact that she remembered me in this special way was totally unexpected. I have those paintings to this day and I cherish them, but the most surprising thing of this commentary, is the fact that the attorney who contacted me about my

From One To Infinity With Synergy

legacy was the very same young boy who lived next door to her and to whom I had been introduced to while on Navy leave in 1944. It's funny how unimportant little things can seem to be at the time they take place only to realize much later that the moment was a pivotal one in future events. Maybe there is a lesson here for us to consider. I personally believe that every moment is precious, and that it should be reflected on as important to the extreme. Now that I am in my autumn years, I realize how true this is.

From One To Infinity With Synergy

The Death Of My Father and Taking The Blame.

When my father passed away, I was lost. The finality of his passing was something that I couldn't quite fathom. It's as if I expected that something would happen to change what I was experiencing, and somehow he would be back in my life as if nothing had happened. It's hard to accept that a person who is so much a part of you, will never be again. For a youngster, it's the height of denial. To begin with, it's incomprehensible. But I remember clearly that morning that my mother had gone to the hospital to be with my dad and I decided that I would go to the hospital myself. The hospital was very close to where we lived and in no time I was there. I arrived to find my mother crying and when she saw me, she grabbed me and held me tight as she said, "Wallace, your daddy is dead, kiss your daddy goodbye." In my mind's eye, I can see him just

From One To Infinity With Synergy

as plain as if he were before me right now, lying on the bed, his eyes closed, and very still. The main thing I noticed was that he wasn't breathing. He just laid there, still. I touched him and felt his warmth. He had just died before I came into the room. I bent over and kissed him as my mother had told me to do. Getting no response from him, it was then that the reality of it hit me, and I knew he was gone.

That wasn't my first experience with someone who had passed on, but the previous occasion was back in Havana, Cuba with the passing of my grandmother, and although I loved her dearly, my loss didn't compare to the passing of my father. My father dearly loved all three of us, but for some reason, I've always felt that I was special to him, which I now understand is not uncommon thinking. For a long while, my dad not being around just didn't fit in. We take people for granted, never giving thought to the possibility of their not being around. He was always there when I needed him, and to accept the fact that his love and counseling would no longer be available to me was just something that I couldn't accept. I was in total denial, and it affected me to the core of my being.

From One To Infinity With Synergy

Things would flash through my mind about him. I recall, as a child while living in Cuba, he would depart our home and I would follow him to the front yard. There I would look at him as he walked to the refinery grounds and to work. I would yell at him "Bye Daddy" "Bye Daddy" and he would turn around every time and yell back "Bye Son" I don't know how many times I would yell at him saying my good byes and he always turned and called back to me. This would go on till he was just a small dot in the distance and could no longer hear me. This was a ritual that took place every morning and I can see it plainly now as if it happened yesterday. Another thing that stands out in my memory of him was something of a ritual in the home. I was always getting into his lap, especially when he was trying to read his paper. Although he had hair on the sides and on the back of his head, he was somewhat bald on the top of his head. I also remember that he had very large veins on his scalp which fascinated me. I would run my index finger along the veins which were so prominent on his bald head pushing the blood along with my finger, and it would be like I was in a car on a road which was weaving back and forth. For some reason, I got great pleasure doing that and would giggle at the fun of it. I know I was a bother to him, as he tried to read his paper, but not once did he ever complain. He really was a patient man.

It's funny how you recall certain things. When it begin to sink in that my dad was really gone and would never return, I started to wonder to what degree I was responsible for his not being there. As a grown adult, I now know that children quite often blame themselves for the loss of a father or mother. I was no different at the time. I would recall, that my father had a saying "You kids are going to be the death of me yet." This statement was an aside to some activity his kids were up to which he didn't approve of, and it was said due to something I had done on more than one occasion. It just goes to show you how important little things can add up to. It's natural as I said, for a youngster to try to figure out if the death of a parent was in any way something he might be responsible for. Those words of his came ringing back to me at that time, and it bothered me something awful. I just knew that I had something to do with his death and was somehow responsible for it. It was very difficult dealing with that guilt, and I am sure that it affected me at the time in ways

From One To Infinity With Synergy

emotional. Those days were very difficult for me, and as I said before, I really don't know how I would have made it had it not been for Mr. Duggan, the kind gentleman who looked after the family, Ms. Smith, my home room teacher, and Ms. Norris, who was my music teacher. There were many more, but they stick out prominently right now, as the people who most were responsible for my transition as a young juvenile into the stark world of reality. They helped me grow up fast. I began to see my world the way it really was. The year 1935 was a rough year not only for me, but for countless others. I saw I was living in a world capable of boundless abundance, and yet poverty surrounded us all. The depression was hitting us hard.

From One To Infinity With Synergy

Western Union to the Rescue

I remember I was 14 at the time, and I was having a running conversation with myself about the human condition and the role I was playing in it. Suffice it to say, I wasn't very happy with things overall. The Depression was in full swing, my mother was working for the W.P.A. (Work Progress Administration) as a seamstress, I was hawking my Extra's whenever they came out, and with my shoe shine box, I was picking up a few cents here and there. But things were rough and I wanted to contribute more to the household needs, so I decided I had to get a job!

I had heard from some of the newsboys, that Western Union needed some Messenger Boys. I immediately got myself to the main office and got an interview with a nice gentleman who asked me all kinds of questions. First off, how old I was, well I knew you had to be 16 to even be considered, so naturally, I lied! Stood right there in front of that man and told a bald face lie. Although I don't think I had hit 110 pounds yet, he seemed to accept the fact and we went on with the interview. The next question, I was really fretting over. I knew it was coming, and it did. Looking me straight in the eye he said "Do you have a bicycle?" Without blinking an eye, I lied again! "Oh sure, I said" hoping my lying wouldn't give me away, because the facts of the matter was I didn't own a bicycle. Damn I thought, I'm getting good at this lying stuff. Which really made me uneasy because lying was something my mother wouldn't tolerate and it just wasn't my nature to lie about things. However, I understood that sometimes exceptions to the rule apply if circumstances require it. Those exceptions gave it a name, it was a *white lie*. I figured what I was being questioned about required my answering with a white lie. I had to get that job, so I told a white lie. My mother told me that if you find yourself in a circumstance where the answer to a question might hurt someone's feeling, then perhaps a white lie would serve a better purpose than to hurt one's feelings, but the exceptions were rare and one had to give it much thought. I figured I was encountering one of those rare exceptions, so I told a white lie.

Somehow I must have impressed the gentleman that was interviewing me because the next thing I knew, he was telling me that I could pick up my Western Union Uniform at the end of the next week, and that I was to start the week after that. He did impress upon me the fact that I would have to have my parents sign a form attesting to my age etc. and I took them saying it was no problem and that I would have them back immediately. But you must understand, I was really facing a dilemma. This job required that I own a bicycle and I didn't. So now I really had a problem.

Let me digress for a moment. I have been most fortunate as far as being gainfully employed all my life. It has always been a maxim of mine that I would never be unemployed, because If I ever found myself without a job, in other words, if no-one was interested in hiring me, then by hook or crook, I would make one. I am prepared to convince a potential employer that I would work for free, just long enough to convince him his business would suffer without my contribution. I can't say that I have ever found myself under those circumstances, and you may think its rhetoric on my part, but I merely mention it because I actually would do it if I had to. Another thing that bothers me is the fact that those who are without gainful employment at times just don't seem to realize that when you are out of a job, you still have one, and that is to go out there and find one. No job is going to come knocking on your door. Another thing I have noticed is that if you look for a job, you have to be prepared for the shock of finding one. In other words, you have to be prepared to go to work. Seems to me that sitting at home and drawing unemployment checks doesn't do a thing for the psyche. Now don't get me wrong, I know sometimes things are beyond our control. I can think of nothing worse than to find oneself in a situation where you literally can't find employment and thank God that a mechanism exists for those in that circumstance to be able to hold things together at least for a time with unemployment insurance. Well enough of that, let's get back to Western Union.

Since I didn't have a bicycle, I had to get one, and I had to get it fast. So I put my brain to work, and here is what I did. I knew a gentleman who worked for the School System as a truant Officer, not because I was worthy of his attention, because I was never truant, but because everyone in the Houston School System knew Mr. Foster! He acted as school counselor at times so I knew him in that capacity. I went to him and told him about my recent escapade with the personnel people doing the hiring at Western Union and the fact that If I could just get my hands on a bicycle, I had myself a job. He was very understanding, and would you believe it, he put me in his car and drove me to the Houston Police Department where we met with an officer who was responsible for safekeeping stolen bicycles. They took me into a compartment and there in front of me must have been 50 bikes of all sizes and descriptions. He told me to pick out any one I wanted, which I did, then they had a person type a letter with the Houston Police Department letterhead on it, describing me, and the bicycle in detail explaining that if the owner of the bicycle ever encountered me accusing me of theft, that all I needed to do was produce the document which explained why I had the stolen bicycle. I rode the bicycle home feeling like I owned the world, when I realized that when I got home there would be another problem I had to hurdle, and that was my mother. However, I knew I had a guardian angel I could count on that had never failed me up to this point, so I went home with some degree of confidence that somehow things would work out.

From One To Infinity With Synergy

Western Union & A Near Death Experience

As you know, through the help of Mr. Taylor, I was able to get my hands on a bicycle. I got my Western Union uniform, including the leather puttees that all messengers wore in those days, and went to work.

I was 14 at the time, and I can assure you, I knew the city of Houston like the back of my hand. Besides that, the main office I worked out of had large maps and street address locators which were available to us. If we were given a telegram to deliver, we would punch a time clock and the location of the address would be identified by concentric circles with the main office at the center. Each band was identified in alphabetical order with so many minutes allocated for the delivery of the telegram dependent on what alphabetical letter the address was located. With all good luck locating your party, and depending on just how hard you "pumped" to get there and back, it wasn't uncommon to get back to the office with some time to spare. I was determined to be the fastest deliverer of telegrams in that office. In no time, I showed my employer that I could be depended on to deliver my telegrams with speed and punctuality.

There are kinds of tricks one learns fast when one is working on a job. As you become familiar, you learn those things that work and those don't work as well. Being young, and energetic, I was determined to learn fast, and I did. For instance, since time was of the essence in the delivery system, you were constantly aware of the time which you were allocated for the particular run you were on. The idea was to punch that time clock when you returned with time to spare. If you were given 30 minutes from stamping out to stamping in, you did everything you could to beat the clock. That messenger boy who slowly built up those precious minutes by getting back sooner than expected, was looked upon with favor. I was one of them. However, let me tell you how I did it.

From One To Infinity With Synergy

Western Union had a policy that everyone had to comply with, without exception. Under no circumstances were messengers allowed to hitch a ride, by grabbing the side or rear flatbed of a truck for instance, and getting a "free ride" as it were. Besides the fact that it was against the law, it was dangerous as well. The bicycle I had chosen at the police station to start my job with, had handle bars that were long and beautiful, sticking out to each side like the horns on a Texas longhorn steer. That meant that I had to carefully calculate how much space I had to weave in and out of traffic when I was in a hurry. Naturally, in no time I was doing maneuvers with that bicycle like a pro, and in the interest of maintaining my reputation for being the messenger boy who was delivering telegrams faster than anyone else, I started getting careless.

Let me digress for a moment, when my father was running the Texaco filling station he owned in Houston Texas, a motorcycle police officer would stop by and have a coke to refresh himself. He was a friendly guy, who would always answer the questions put to him by those youngsters who are naturally drawn to the persona an officer of the law presents,

From One To Infinity With Synergy

especially if he rides a motorcycle. We would ask questions about his Harley Davidson, what it was like to chase someone at breakneck speeds if they were trying to elude him, and he would tell us of some recent episode about just that. As kids we stood in awe at this local hero and I remember telling him that someday I would own my own motorcycle, because they always intrigued me. Realizing what I had said might come to pass, he proceeded to tell us all that motorcycles were OK, but that they were dangerous. The problem he said, was, that the more you ride them, the more familiar you are with them and the more you think you can do with them, and so you start throwing them around doing foolish things. The next thing you know, they are throwing you, and you are in deep trouble. It was just a passing remark on his part, but I never forgot what he said nor did I forget the name on his name badge, it was the same as mine, W. Johnson. A week later I learned that he had been killed riding his motorcycle while rounding a curve. I don't know under what circumstances he died. Did he try to throw his Harley around just one more time and the monster he was riding rebelled and threw him instead with its subsequent tragic ending? I never found out, all I

From One To Infinity With Synergy

know is I never bought my own motorcycle. In subsequent events in my life, and remembering his admonition, I have often wondered if Officer Johnson ever looked down on me and worried about me and my main fascination with airplanes.

Back to Western Union, since everything revolved around time, you tried to deliver your messages in as short a time period as possible, and being young and reckless, I found myself setting aside the Western Union rule against hitching a ride. After all, by grabbing a truck's side, you could be pulled along at speeds in excess of anything you could do by pedaling. So I found myself breaking the law and hitching a ride. I get cold chills recollecting what happened one afternoon when I grabbed the end of a flatbed truck.

As I said before, my handle bars extended quite a way out to each side and if you hitch a ride you hold on with the left hand and the right hand is controlling the handle bars. Extending further out than our current handle bars do, you have to make sure that they don't come in contact with the truck bed which puts you in an awkward position as you are pulled along.

From One To Infinity With Synergy

Under normal circumstances when the truck you are hitching on gets to a certain speed limit, good judgment and innate intelligence tell you when to let go. For some reason I found myself not letting go, not only that, I could see the face of the truck driver in the rear view mirror looking back at me and laughing as he had slowly increased his speed past the point of no return. Now I was going at a speed where I was afraid to let go and he knew it. I was hoping traffic or a light would force him to slow down, but instead fate was against me, and the longer I held on the more he increased his speed. I realized I was in a dangerous situation, if I didn't let go, I was going to get killed for sure, and I was scared to death. I summoned all the courage in me and decided to let go hoping that I would be able to get my left arm back to the handle bar so as to control my bicycle. The Gods were with me that day, and after stopping for a short while, I realized how close I came to getting into real trouble. There is a theory that says that there is no such thing as reality, but only perception. Like the cat with nine lives, for all I know I could have been killed, but I perceived otherwise and carried on. I know one thing for sure, I was scared to death and back.

From One To Infinity With Synergy

I never hitched a ride again ever, but the time for me to deliver a telegram took just a little more time than my overall averages indicated I should take, and my boss never new why and I wasn't about to tell him.

From One To Infinity With Synergy

Was I A Bag Lady For the Mafia?

You know, there are cigars and then there are cigars! I mean REAL cigars. I have never smoked, but I did go through a pipe smoking phase way back in 1946 that lasted three months I think. The darn thing was more trouble than it was worth, and like one of our presidents, I never inhaled either. All I have to do is put a cigarette in my mouth and without inhaling at all, I start levitating and in no time I'm higher than a kite in a brisk wind. Now, why am I digressing from my journal which on the last account was way back in 1939?

Well let me explain. As you know, I grew up as a young boy in Havana, Cuba. As I was saying, there are cigars and then there are REAL cigars. I'm not a smoker, but I know a real cigar when the aroma of one comes across my sensitive olfactory nerves. And unless you have had someone light up a real Cuban, hand rolled cigar, and caught a whiff of it, you won't know what I'm talking about. Just ask any current resident of Miami who is a Cuban and he will immediately tell you that the only cigar that is a *real* cigar is one that is hand rolled from leaves that are grown on the Island of Cuba. And although illegal to currently bring to the states, it is said that our own president Kennedy, who knew a good cigar, would on occasion fill the oval office of the White House with the aroma of his Cuban cigars. I wonder how he got them. But again, I digress.

Delivering my newspapers in Houston, Texas gave me the opportunity to meet some rather interesting characters. One was a swarthy plump gentleman to whom I would personally hand the newspaper I delivered at his address. Sitting in front of his business establishment, a used car lot, he would rock back and forth in his rocking chair every afternoon waiting for me to arrive, all the while blowing smoke rings that emanated from the biggest cigar I have ever seen in my life. Not only was it big, you could tell that it was not a cigar that came from some assembly line by the hundreds, but rather a cigar that was hand rolled for sure. A perfect

From One To Infinity With Synergy

example of a hand -made Cuban cigar. There is nothing worse than second hand smoke as far as I'm concerned. Smoking is a filthy habit I'm glad I didn't get hooked on. But, I have to admit, provided that it doesn't hit you full force, there is something about a Cuban cigar that's different. And l can understand why there are private haunts where gentlemen meet to this day in their exclusive clubs with their own private humidors which contain of all things illegal Cuban cigars. They say rank has its privilege and so does money, which I guess explains a lot, from President Kennedy on down. Anyway, I would hand him the paper and he would blow me a perfect smoke ring. Quid Pro Quo I guess. This was a routine that followed day after day come rain or shine. He seemed like a nice enough guy but all the time I sort of felt that he was just a little different. I wasn't too worldly for my age when it came to know human character perhaps, but I wasn't stupid either, and for some reason I just didn't feel comfortable when I was in his presence. He was too slick I thought. And now that I look back on it, he reminds me of that old character actor Edward G. Robinson, voice and all. But I was too young to fully grasp the meaning of the whole thing, and on one particular day, he asked me

From One To Infinity With Synergy

to come back to his place of business, because he wanted me to do him a favor. Upon completion of my appointed rounds with the newspaper route, I returned to him as he asked.

The favor he was asking was that I take a package about the size of a shoe box to an address in an area of Houston which was crime ridden and harbored some rather unsavory characters. My instructions were to deliver the package to a person in the hotel, whereupon he was to call the cigar smoker that I had arrived with the package, and he then would then pay me $5.00. Now $5.00 in 1939 was big money and although at the time I was too naïve to realize that something wasn't just right, accepted the money gladly, but I knew that the people who lived in this hotel weren't exactly the kind of people I should have anything to do with. There were Chinese and Latino men there and it smelled to high heaven of marijuana and liquor, accompanied by a pungent odor which I have only smelled in the orient which I now believe was opium. Suffice it to say, that when the cigar smoker asked my once again to be his errand boy, I declined saying I was told by my parents not to go into that part of Houston, he didn't press the issue, and I was glad.

From One To Infinity With Synergy

That $5.00 was easy money to come by, but like all easy money, there were strings attached, and I wonder to this day just what was in that box. I now know I was being used as a bag lady. Was it narcotics I was transporting? Or cash for dope, I don't know, but in retrospect, I shudder at the thought of the real mess I could have been getting myself into. I know this; whatever was in that box was illegal as Hell. I want to think that it might have been a box of Cuban cigars, but I know better. You see, I know what a Real Cuban cigar smells like.

From One To Infinity With Synergy

Hawking "Extras" & Getting an Education

Although I acquired my social conscious early on, it was by selling newspapers that I was really able to fathom just what was really going on in the world. The thirties were filled with daily happenings from all around the world. The depression was in full stride in the states and Stalin, Hitler, and Mussolini, were being given credit for "Running the trains on time." which was an ominous statement when you really think about it. Wars and rumors of wars were rampant, and any time anything of any consequence occurred, the newspapers would come out with a special edition called an "Extra." I remember on more than one occasion being up early in the morning and going to the "Houston Chronicle" and picking up 15 or 20 newspapers, jumping on my bicycle and yelling at the top of my voice "Extra, Extra" and whatever was the heading in bold letters

From One To Infinity With Synergy

on the front page, which was the reason for the extra publication in the first place. Just by reading those headlines and a little of the byline, I pretty well got the gist of the cataclysmic events that were daily occurrences, not only in the U.S. but particularly in Europe. As I said before, I was really being educated rapidly and at an early age about things which we can reflect back on and be proud of, while at the same time getting a sick feeling in the pit of one's stomach at the sheer stupidity and blunders perpetrated on us by so-called leaders of the world. Besides hawking my "Extras" of course I had a paper route. So every day I would go to the pickup point and load the bags which were hung on each side of the rear wheel of my bicycle. There were three major newspapers in Houston at the time, The Houston Chronicle, which had the largest circulation, followed by the Post, and then the Press. On Sundays, I would pick up my allotted number from the Chronicle and the Post and hawk them on Saturday nights as an early Sunday morning edition. I have nostalgic memories of going into a ratskeller near the "Uptown" Theater reeking with the aroma of cigarettes, cigar smoke, and beer, and at the same time hearing the melodious voice of

From One To Infinity With Synergy

Bing Crosby coming out of the Wurlitzer Juke Box singing that beautiful song *Sweet Leilani*. To this day, it is *deja vu* all over again, a saying made famous and attributed to Yogi Bera, the catcher for the Brooklyn Dodgers, if I remember right. Who can remember Neville Chamberlain with his famous umbrella, returning from his visit with Hitler and uttering those famous words which would haunt him to his dying day, "Peace In Our Time". What were you doing on Sept 3rd. 1939 at 3:00 A.M. in the morning? You can't remember can you, well I can, because I was smart enough to know that WWII was just about to start, and I was at the dock of the Houston Chronicle waiting for the "Extra" to start rolling off of the presses with the announcement that England had declared war on Germany. And sure enough by 6:00 A.M. that morning I can proudly say I sold papers like they were going out of style, yelling at the top of my voice "Extra-Extra England Declares War." Not in downtown Houston, but in the suburb where I lived. From my shrieking that headline, I bet I awoke more than one person from a sound sleep thinking that the world had ended or something as or something as calamitous had happened, and of course it had.

From One To Infinity With Synergy

The US Navy and The Banana Caper. Where There Is A Will, There Is A Way.

I graduated from Dow Jr. High and enrolled in Sam Houston High School which was downtown and was accepted into the Debating Society and into the School Band. Previously I played the Cello and Violin, but this time was assigned the Base Viol. I didn't mind as long as I could participate in the music program. High school was a complete change scholastically, but I applied myself and in short order was making friends and doing well with my studies. Unfortunately due to circumstances beyond my control, I soon found myself in a pickle.

Houston has its schools assigned by wards, you had to go to the school in the ward where you lived. Mother was forced to move into a ward that

From One To Infinity With Synergy

required I go to San Jacinto High School and that meant I had to change schools. Up to that time, I had never had a problem with social Interaction with my peers. At Dow and Sam Houston, my friends and their families were in the same financial circumstances as I was experiencing. My clothing attire was in keeping with the rest of my peers and my mother saw to it that I was always neat in appearance, although I did have in my prized possession, a pair of black and white wing tip shoes, and a too large hand me down suit given me by my uncle. I wore the whole get up for special occasions such as music recitals etc. Regardless, it was obviously not tailored to me. But in those depression years, people made do, and I wasn't the only one wearing hand me downs. People, especially the teachers, understood.

San Jacinto High and Sam Houston High were rivals in all things. Students at San Jacinto were saddled with the pejorative term "The Tea Sippers" due mostly to the fact that the school was in a Middle Class ward where the families were much better off financially. This was reflected with students who drove their own cars, and wore seersucker slacks. On special occasions, students'

From One To Infinity With Synergy

shirts had cuff links. I was always neat, but the best I could do for special occasions, was my hand me down suit. I was a site for sore eyes, and I knew it. But not all students lucky enough to live a high middle income life style are pretentious snobs, and in short order I was included in the "Inner Circle" as one of the guys. As a Sophomore, I befriended one student who was a Senior. He was in the debating class with me and was one of the students who drove his own car. As an aside: (He became an extremely successful lawyer in Houston after WWII. If I were to mention his name, you would recognize it immediately because he was nationally known). But get this picture. He would invite our small group to pile into his car and go to Prince's Drive In where short skirted car hops would roller skate out to the car with trays that attached to the windows. On the trays would be hamburgers, malted milks and cokes, with abundant potato chips. At the time, I was working for Western Union after school hours and on weekends, and I always had a few pennies in my pocket, but as was generally the case, not enough. I could go the malt, or the hamburger, but not both. My friends knew this and quite often I would be the recipient of a treat. But I could never accept without a sense of guilt or that I was somehow mooching. It made me uncomfortable to say the least. I knew it had to change.

On September 3rd. at 4 AM, I was waiting for the Houston Press to release what anyone with any grey cells in their brains, knew that war was coming. Neville Chamberlain the British Prime Minister with his famous umbrella had just returned from Europe to proclaim that "We Will Have Peace In Our Time." This, after getting a promise from Adolf Hitler, that he would not invade Poland. Shortly after, Hitler broke his promise, and we headed rapidly into hostility. The rest is history, but for me, it was a momentous occasion. I was given 100 of the extras "England Declares War." and off I went to hawk those extras which sold for a dime and which I made four cents from. Four dollars for a kid 16 years old was a lot of money, that and the twelve cents an hour I was paid as a Western Union Messenger Boy kept me in spare change, but not enough to keep up with my "Tea Sipper Friends." As I said, something had to change, and so I set out to bring that change about.

From One To Infinity With Synergy

That summer, I had tried to join the CCC's. (The Civilian Conservation Corp." but I was too skinny and frail to pass the physical. Reflecting on it I now realize how lucky I was they didn't accept me. Those guys in the three C's were given hard work to do. Fighting forest fires among other things comes to mind. I would not have lived through it for sure. So I gave that up and decided the answer was to enlist in the military, and I chose the Navy.

I remember talking to the recruiter, a Navy Chief Petty Officer (which later I would become myself), and told him that I was 17 years of age. He responded that I needed permission from my mother and a statement authenticating my date of birth. I rushed home, and of course she refused. I remember getting on my knees with my hands clasped in sublimation pleading for her to sign. I cried out to her, and explained how unhappy I found myself to be, and she cried along with me. She had experienced sadness and grief as well and saw I was sincere in my pleas and why I was making them, so I prevailed and she signed the note I had written.

The recruiter showed surprise when I returned in such short order and I now think he never expected to see me again, but he accepted the signed statement, put me on a scale and said that I could not pass the physical because I was too skinny. For my age and height I had to weigh in at least at 111 pounds, and I only weighed 110. I asked if he could make an exception, and he said "No Exceptions." He replied "come back in six months and if you weigh 111 pounds you will qualify." I left the recruiter pondering what to do. I am absolutely sure now, that he was pulling my leg all along and was just trying to get rid of me. He probably guessed I was not in fact 17 and as for the weight thing, he used that as a ruse to get rid of what I am sure to him, was an annoying kid.

Not far from the recruiting office, there was a farmers market. I would pass it every day going to and from while attending San Houston High, and would on occasion, buy a single banana. While living in Cuba as a child, bananas were a daily staple, inexpensive and abundant, and I happened to like bananas. I came upon a plan to pull one over on the chief recruiter. I bought what must have been at least a couple of pounds of bananas and preceded eating them in short order. It took some time, but I got them all in and promptly returned to the recruiter. The look on his face told me he was surprised to see me, and I said to him, "If I weigh 111 pounds, will you take me?" He said sure, and when I stepped on the same scale that I had used only a few short hours back, with the same clothing attire and no visible means of cheating on my weight, I weighed in at just over 111 pounds. He looked at me with a bewildered look on him face, smiled and said, 'You are in, kid."

On September 30th. 1941, I was sworn in to the service of the US Navy having pulled off the "The Banana Caper." To this day, I still like bananas.

From One To Infinity With Synergy

THE MILITARY YEARS

From One To Infinity With Synergy

It Was a Dark and Stormy Night

It was a dark and stormy night is an often-mocked and parodied phrase written by English novelist Edward Bulwer-Lytton in the opening sentence of his 1830 novel Paul Clifford. The phrase is considered to represent "the archetypal example of a florid, melodramatic style of fiction writing," also known as purple prose. Nevertheless, he became famous by using the phrase in the first sentence of his novel. I'm not trying become famous or anything of the sort, and although his novel was fiction, I assure you this is an actual event I played a small part in. Here are the facts as I remember them.

It *was* a dark and stormy night. I was on the bridge of the USS Jamestown AGP-3 (Motor Torpedo

From One To Infinity With Synergy

Boat Tender), as the duty signalman on the night of October 17, 1945. The ship was laboring under heavy sea conditions, visibility was low with a low overcast as well. Lt. K. W. Prescott who was our Executive Officer, was Officer Of The Deck, and except for the extremely bad weather, everything was routine.

We were underway heading for Borneo if my memory serves me right, when all of a sudden out of no-where, a Billy Mitchell B-25 flew right over the ship. He immediately made a steep turn and returned to pass over us once again at extremely low altitude. The captain was informed and we all realized that the planes actions clearly indicated he had no desire to leave us. All attempts at radio communications were to no avail. If a plane was ever in trouble, it didn't take much thinking to agree this one had a problem. The question was, what to do?

We were operating under complete blackout conditions, and there wasn't much we could do. But we decided to try and help the stricken plane by visual means. Among the many things which we as signalmen used for communications between ships

From One To Infinity With Synergy

ie. Flag Hoist, Semaphore Flags, was a devise called a Signal Gun. It was about two feet long, and about four inches in diameter, and it contained a very bright light bulb. It was basically a rifle that emitted a very narrow beam of light. Pointed directly at the receiver of the light, it made it possible under blackout conditions to use a bright light at the exclusion of anyone other than the person it was pointed at. In that way, you had a degree of security from disclosing your presence to any possible enemy which might be in the area. It worked well under normal conditions, but if you weren't exactly on the target, the narrow beam would make it difficult for the recipient to see it. Adding to the physical restrictions of the Signal Gun under proper use, throw in the fact that you are trying to fire the light at the cockpit of a fast moving plane. On top of that, we were in heavy seas with the ship rolling and pitching intensely, which made matters worse. The Captain ordered me to send the magnetic heading from our position at the time, to the nearest port of Zamboanga, Mindanao. I know that part of the flight training given by the military is the requirement to be able to read the International Morse code. I am positive that the pilot of that ill-fated plane, 1st. Lt. Austin C. Fitzgerald U.S. Marine Corp. was really glad he paid attention during his flight training class when he was taught the Morse Code.

I very slowly but surely sent the Morse Code of the magnetic heading to Zamboanga. I repeated it over and over. All the time, the plane kept circling and circling the ship. It was decided that another visual aid might be of aid. The captain ordered we turn the ship almost 270 degrees and take up the magnetic heading to Zamboanga. That pilot immediately realized that we were using the ship as a pointer and he flew the plane directly over the center-line of the ship and disappeared into the dark horizon ahead.

Many things transpired during my tour of duty on the Jamestown during the four years I was a crew member that caused me worry, but nothing compared to my worrying for the crew of that plane. The weather was terrible and I knew they were in trouble. I didn't hold much hope for them.

Imagine our surprise a couple of days later, while

From One To Infinity With Synergy

at anchor in Borneo. when a B-25 Billy Mitchell came out of the blue and started buzzing the hell out of us. We knew immediately that it was the same Billy Mitchell B-25 that we had seen a few nights earlier. It was evident by the actions of that bird that it was a happy one. Let me explain something. You have to understand, since I am a pilot myself, that airmen are a breed apart from the rest of humanity. First of all, when you fly a plane, the pilot becomes a part of the plane as he maneuvers it through the air. It's as if the mass of metal turns into a living thing responding as a living entity responding to the subtle kin-esthetic inputs of the pilot. It isn't a case of a plane with a pilot in it, nor is it a pilot in an airplane. The plane becomes the pilot and the pilot is the plane It's a metaphysical thing that only pilots understand. That plane was showing its exuberance at being alive as well as the crew was. I cried inside myself with the joy of knowing they had made it after all.

I have often wondered what ever happened to those valiant and brave airmen since that fateful night in October 1945, and I prayed they made it to the end of the war safe and sound. I have always felt bad about the fact that I didn't follow up in trying to locate that pilot and crew, but at a subsequent re-union of the USS Jamestown, Lt. Prescott (Now Captain USNR) who was the officer of the deck that night was quite surprised when I showed him one of my WWII mementos. I am duplicating them verbatim for all of you to share with me.

USS JAMESTOWN (AGP-3)

AGP-33/F15

28 October 1945

From: The Executive Officer

To: JOHNSON, Wallace Albert, SM1c USN 360-47-98

Subject: Letters from Port Director Zamboanga, Mindanao

Dated 21 October 1945.

From One To Infinity With Synergy

1.   The subject letters are forwarded to you in a much as this command believes that your performance as duty signalman the night of October 21st. was instrumental in giving the pilot his "steer home", and should therefore give you great pleasure in the realization of a job well done.

K.W. PRESCOTT, Lieut., USNR

21 October 1945

To: The Commanding Officer, U.S.S. Jamestown AGP-3.

1.   The enclosed message was left with us for delivery to you after your departure last night. We were unable to affect delivery because you were out of voice range and therefore take this means of doing so.

Lt. (jg), MAC BAIN SMITH Office of Director, Zamboanga, Port Mindanao

From One To Infinity With Synergy

On Wednesday October 17, 1945, while flying on a routine local hop, all of my Radio and Radar gear burned out and the weather closed in completely in this area. With 2 hours Gas left and no idea where I was, I fortunately sighted you and was able to get a steer home. On behalf of my crew, we tried to contact you to have you all over for dinner, but missed you twice. If you are ever in this area again please come up to Marine Bombing Squadron 611. There is nothing we won't do for you. God Bless you all.

AUSTIN C. FIRZGERALD, 1st Lieutenant, U.S. MARINE CORP

From One To Infinity With Synergy

Surprise Inspections and the Roving Still

It was always a wonder, not only to me, but the rest of the crew. We would leave port, be at sea for a couple of days, and all of a sudden, a disturbance would take place on ship; usually a fight between two who were obviously drunk. It would end up with the Master at Arms putting them in the brig, followed by a Captains Mast finding them Drunk and Disorderly with a sentence of so many days in the pokey on piss and punk (Bread and Water). Not only that, but it was always the same people, just a handful of course, but you could count on it sure as all get out. The question always was the same: Where did the booze come from? When asked by the Captain, they would naturally say nothing. So from time to time, unknown as to when it would happen, a surprise inspection would take place. But first, let me give you a little history.

Naturally, drinking liquor on board ship was against Naval Regulations. That is not to say we didn't have beer aboard. A supply ship would send over what the Jamestown would be entitled to, and we would have a working party bring it on board and stored in the storage area. Say we were entitled to 400 cases. An officer with a tally sheet would take inventory as it left the supply ship and put on a barge for transport alongside the Jamestown, then again, the inventory count would take place as it came aboard, each case taken by one of the working party, and transported to the storage area when again an officer with a tally sheet would count the cases and store them. But it wasn't uncommon, and it happened more than once, that the last case would come on board, but when it arrived at the storage area, the tally sheet of the inventory was never 400 cases. There was always be a shortage of one or two cases. The officer on the barge couldn't explain it, nor could the officer at the end of the trail in the storage area. Naturally, the Captain was pissed off to say the least. But no one would own up. Well the two drunkards would never admit to the heist of the beer, but it would explain why they were drunk. However it was a mystery never solved.

We would from time to time be allowed to go ashore for rest and recreation and a ration of three cans of beer were allotted for each crew member. Those of the crew who didn't drink, and there weren't many, were still entitled to their three cans. That opened up the opportunity to barter the three cans for cigarettes perhaps, or to outright sell them for cash. But it did make it possible for someone to consume more than his allotted three cans. That being the case, the recreation party always came back to the ship with someone who was obviously under the weather as it were. But his being drunk could be explained. Consuming more than three cans was not looked upon kindly and they tried to control it, but someone always came back drunk. Always the same people were involved. You have to realize, the crew was made up of all kinds of people from all walks of life and I am sure we had people on board who were heavy drinkers long before they arrived on the Jamestown. I have seen some guys who actually drank Shaving Lotion for its alcohol content. I am sure it wasn't a problem that only the Jamestown had to put up with.

From One To Infinity With Synergy

Coming back to the ship drunk after a recreation party beer bust was understood by the command, but being at sea for a few days, and all of a sudden we have a drunken brawl was disconcerting to the Captain to say the least. What to do? The answer: *A Surprise Inspection.*

He would start at the bow of the ship and work his way aft. Looking at every nook and corner for the stash he knew was on board someplace. And he wasn't looking for beer, he knew there was a still brewing away somewhere on his ship and he was going to find it.

What the captain didn't know was, sure enough, there was a still brewing away. Don't forget, we were a Motor Torpedo Boat Tender with torpedoes that we supplied the torpedo boats with. They were driven by a propulsion system that used 140 proof alcohol, but it was under tight control, so he didn't suspect it came from that source. But what he didn't know was the resourcefulness of the guys who did have a still brewing away. As an example, we would from time to time have apricots or prunes, which was printed on the menu of the meals the commissary steward

From One To Infinity With Synergy

was required to publish on a weekly basis. You have no idea how much these guys were willing pay for your ration of prunes. Get enough prunes together and under proper care you can come up with some real Kickapoo Juice. Just about any fruit will ferment into alcohol if you know what you are doing. And these guys were good, so the still was on board alright, but where? The Captain would scour the ship from stem to stern, and he did this on more than one occasion. But he NEVER found the still. The question always on my mind, and I'm sure it drove the Captain crazy as well was, how did they maneuver the still around him as he performed his Surprise Inspection of the ship? No one would ever admit to having the still, and the Captain *never* found it. Another mystery at sea never solved.

I was on the USS Jamestown over 4 years, and it never failed that after leaving port and being at sea for a few days, someone would show up drunk. Don't ask me where the still was, I didn't drink then, and drink very little now, but those drunkards on ship made life interesting from time to time.

From One To Infinity With Synergy

THE APOLLO PROGRAM

Life Changing Event III, the 1960s: The Apollo Project

I don't know who said it, but some wise sage is credited with saying that in one's lifetime, one goes through five important life changing events. I have given that some thought and I have decided to make those events the five major entry points for my blog. I have put them down as follows:

Event I. GROWING UP IN HAVANA, CUBA, 1925 to 1932

Event II. RETURNING TO USA. THE DEPRESSION YEARS. 1932 to 1941.

From One To Infinity With Synergy

Event III. RETIRING FROM THE MILITARY AND JOINING NORTH AMERICAN AVIATION. The APOLLO PROGRAM. 1960 to 1970.

Event IV. JOINING LITTON INDUSTRIES 1973 to RETIREMENT 1992.

Event V. CURRENT: RETIRED, LIVING THE GOOD LIFE

When I hired on with North American Aviation, I was put on contract to the Strategic Air Command. I had a Top Secret Clearance and was acting as a civilian in-flight Inertial Navigation Instructor flying in B52's out of Columbus AFT, Mississippi. My duties taxed me physically and mentally. It was a very important job and I was proud to have been chosen to perform it. Keep in mind, although retired military, I was still considered a civilian, and I was flying 13 hour missions that covered much of the globe's geography requiring in-air refueling from KC135 tankers. All the while, the B52 had live nuclear weapons on board, and our training missions would take us from Mississippi to Chicago, Ill, where

From One To Infinity With Synergy

we would in a simulated strike, destroy the city of Chicago. We would then fly to Miami, destroy it, then fly out to the Gulf Of Mexico, refuel, fly to San Francisco, destroy it, then Seattle, followed by In-air refueling again, destroy Houston, then go back home to Columbus AFB, Mississippi. It was a responsible job that needed to be done, and I was qualified to fill that role. But the realization that should some adversary decide to attack us with nuclear weapons, finding us at war, knowing that the attack profile under those circumstances with a B52 which I was flying in, with a certainty of 100%, that I would never return home, made me realize that at any moment I could be experiencing a Life Changing Event for sure.

Imagine how pleased I was to realize that the Company I worked for had won the Apollo Contract. I can still hear President Kennedy making his speech to the Congress and saying "First, I believe that this nation should commit itself to achieving the goal, before this decade is out, of landing a man on the Moon and returning him safely to the earth. No single space project in this period will be more impressive to mankind or more important for the long-range exploration of space; and none will be so difficult or expensive to accomplish." Imagine my mind racing at the thought. As important as my present job was, what could compare with the possibility of being involved with such a noble endeavor. I immediately went about the business of finding out just how I could somehow get involved with the program. I put in for vacation and flew out to the West Coast where my home was in Long Beach, Ca. just a few miles from Downey, California, and the home of North American Aviation. I ultimately met Dr. Joel Canby who was the department head of the Human Factors Group who were working on the Apollo Program. It took me about four months, but ultimately, Dr. Canby had me assigned to his group and I transferred back to Downey and started the most amazing Event of my life up to that point.

I had the pleasure of meeting and closely working with the "Original Seven Apollo Mission Astronauts" chosen for the Lunar Landing Mission.

Group members
Alan Bartlett Shepard Jr., USN, (1923-1998)
MR-3 (Freedom 7), Apollo 14

From One To Infinity With Synergy

Virgil Ivan (Gus) Grissom, USAF, (1926-1967)
MR-4 (Liberty Bell 7), Gemini 3, Apollo 1

John Herschel Glenn Jr., USMC, (born 1921)
MA-6 (Friendship 7), STS-95

Malcolm Scott Carpenter, USN, (born 1925)
MA-7 (Aurora 7)

Walter Marty (Wally) Schirra Jr., USN, (1923-2007) MA-8 (Sigma 7), Gemini 6A, Apollo 7

Leroy Gordon Cooper Jr., USAF, (1927-2004)
MA-9 (Faith 7), Gemini 5

Donald Kent (Deke) Slayton, USAF, (1924-1993) Apollo-Soyuz Test Project

All of these men were very close to my age at the time and now that I am 85 myself, many of them have made the transition which we all will eventually make, leaving these earthy bounds and becoming once and for all part of the Cosmic realm.

From One To Infinity With Synergy

The Tragic Fire of Apollo 1

I had been involved in writing the procedures for removing the double hatch required for extra vehicular activity. At first the NASA insisted on an outside hatch opening to the space environment. The inner hatch would have to be removed inwardly into the command module after decompression allowing the vacuum of space into the capsule. The NASA wanted this double hatch concept because it offered a sense of redundancy in case the outer hatch experienced some kind of pressure failure. They figured correctly that the internal pressure of the capsule would be a pressure against the inner hatch which would insure the hatch would not fail with a leak. They were correct of course but our engineers were of the opinion that the single hatch would offer sufficient safety to circumvent a

From One To Infinity With Synergy

decompression failure to the capsule. NASA won the argument. When deciding to have extra vehicular activity, the capsule had to be decompressed. Then the inner hatch had to go through the procedure of rotating latches and then bringing the hatch into the capsule and storing it under the center seat. This was the task that Ed White the center seat astronaut had on his hands when the fatal fire broke out on the pad at Cape Canaveral, Florida. Unlike the Russians who had an exotic mixture for their internal atmosphere, ours was 100% oxygen. When you supersaturate any matter with 100% oxygen, It makes little difference what the kindling point is, the result is a ferocious burning and consumption of the material. In short order, due to a spark in the wiring in one of the lower compartment areas, a fire broke out. It was followed by a fast build-up of pressure internally that make it impossible for Dave White to break the inner hatch from its seals. In fact, the internal pressure built up so rapidly that it actually ruptured the capsule. The command module had turned into a pressure cooker. It happened so fast, nothing could be done, with disastrous consequences. I and two other test pilots worked round the clock simulating the procedure and

From One To Infinity With Synergy

capturing it all on film. We were trying to determine what the time-line was to get out of the restraining harness, decompress the capsule, and retrieve the inner hatch. I am in personal possession of the 16MM film given me by North American Aviation on my leaving the company. It is only one of many mementos I have which bring back both sad and happy memories. Ultimately, our engineers won the battle about the single hatch and I was given the responsibility of writing the actual words put on a stick on placard which were ultimately put on the inside of the outer hatch on how to open it in preparation for extra vehicular activity. Talk about synchronicity. I served on the USS Hornet CV12 just before my retiring from the Navy. The Hornet is now here in Alameda, Ca. as a floating museum and believe it or not there is an Apollo capsule on board that was actually picked up by the Hornet. It's a small world, and we never know at what moment we are doing something that we think is mundane and of no consequence. How wrong we are. Every moment is precious and every second of our lives is of paramount importance. Nothing happens by chance. Don't ask me to explain it, but I believe that to be true. There is no explanation for those things metaphysical and I am not about to try and explain them. I merely accept it as a cosmological law.

From One To Infinity With Synergy

Twinkle, Twinkle Little Star

Twinkle, twinkle little star,

how I wonder where you are...

In the early days of Apollo studies we knew that the Lunar Landing Module would separate from the Apollo Command Module and descend to the surface of the moon with two men aboard. The Command Module would remain in orbit around the moon waiting for the return of the Lunar Landing Vehicle. One of the problems we anticipated which caused some anxiety would be the ability to make visual contact one with the other when they would rendezvous and dock. It was decided that some kind of a flashing light on both craft would take care of the matter. But that sounds a lot easier than it sounds. In the first place, was there a time restraint which had to be adhered to, i.e., did we have only so many minutes to correct any miscalculations in the rendezvous such that we had to come in visual contact a soon as possible? If so, what luminosity in candle power would this light require? What color should it be? Is there a color that's better than pure white light? At what rate should this light blink? Is there a frequency that is preferable above all others? All kinds of questions come up related to finding a small object in the blackness of space. What to do?

The answer was found in a Planetarium. North American Aviation leased the facility of the Griffith Park Planetarium where we set up a mock-up of the interior of the Command Module windows. We then were placed in a precise location such that we had a restricted view of the star field which was visible to us out of the windows we were looking out of. Then in the total darkness of the planetarium mixed in and hidden among the star field, a flashing light would start blinking. As a pilot subject on that study, it was my task to find the blinking star and identify its location. We had no idea what the blinking rate would be or its location in the star field. You would think it would be a easy task, but those of us who knowhow to search in total darkness for the slightest thing and are acquainted with what goes on with our eyes had an advantage. For we know that to get maximum capabilities from the use of our eyes at night, we know that you must never focus on the object you are looking for but rather look 10 degrees above, below, to the right, or left of the focus point. There is an explanation to this. There are two cells in the eyeball. Cones which can discern the colors of the spectrum but are not very sensitive to light and therefore of little use at night, and Rods which are color blind but very sensitive. There is only one problem, there are NO RODS in the focal point of the eyeball which is called the Fovea, only Cones are located there, and if there isn't sufficient light to activate the Cones you must rely on the Rods to pick the object up. So prove it to yourself in a real dark room, look directly at the object you're are trying to see and then shift your line of vision about ten degrees and sure enough you will see the object better. If you look directly at it, it may disappear only to reappear if you look slightly off the object. I don't know what the data of that study proved. But I know that whatever they are currently using in blinking frequency, candle power intensity, color etc. is the result of that study. It give me a good feeling to know that I played a small part in it. To me the rhyme *Twinkle, twinkle little star, how I wonder where you are.*

Now you know why.

From One To Infinity With Synergy

CURRENT EVENTS

From One To Infinity With Synergy

Senior Citizens, Computers, and The Internet

For the Senior Citizen in his Autumn years, there are too many tales repeated about how hard computers are to master and how difficult it is to navigate the Internet with them. This is especially true for those who think that they are "Too Old to learn new tricks." Unfortunately, there are too many true stories about scammers who prey on Seniors. Those scammers are just waiting for the innocent wanderer on the cyber highway of the internet to be picked on and fleeced. True, one has to be careful regardless of age, but with proper mentoring and guidance, these pitfalls can be avoided and a senior citizen can easily participate in the technological revolution that is taking place. For the Senior Citizen, the experience can be most rewarding with many benefits to those in their twilight years.

From One To Infinity With Synergy

One of the most common mistakes made, is that you immediately start wondering how you can start making money with your computer. You soon learn that if you have questions, answers are readily available. However, quite often, the answer has a dollar cost to it, and once acquired, the answer leads to more questions, so you fall for it with its ensuing cost, and the next thing you know, you are in a Daisy Chain going round and round, not unlike a dog chasing its tail and getting nowhere fast. In no time, you are completely lost, heading for a Black Hole in the Cosmic Cyberspace Superhighway called the internet.

It is disheartening to say the least, however, it doesn't have to be this way. This article, and those that follow, will let you in on some of the secrets that the author encountered while searching for the solution to making money on the internet.

The most perilous mistake one can make, is trying to do it alone. You must have a mentor of sorts, if for no other reason than to realize that learning by making mistakes is the worst way to learn anything. Another problem, is scattering your shots all over the place, in the hope that you will hit something. You chase after one thing, then another, spending money each way possible, and in the long run, never get anywhere, all because you failed to focus on one thing at a time. Staying focused is difficult, but is of prime importance. The author speaks with experience in this regard and future articles will point out where these pitfalls are and how to avoid them.

There is no reason for thinking that age is a detriment to mastering those skills necessary to be successful. Whether your effort is merely to use the Internet as a hobby, keeping in touch with family via E-mail, or deciding to supplement ones income, all are possible with proper guidance. In future articles, I will endeavor to give you the necessary information and skills, which will give you the confidence to take on the internet without fear of failure. As a former Test Pilot, I will help you navigate through the maze to insure safe passage and the completion of a successful mission.

From One To Infinity With Synergy

A Modern Tragedy

Dear Friends, Since I am a resident of Alameda living on Shore Line Dr. I became aware of the drowning of Raymond Zack in short order. Dr. Jeffrey Lant is an avid article writer who has given his dealers or Worldprofit of which he is the CEO, permission to use his articles in our newsletters and blogs. I immediately notified him about Raymond Zack and he asked me to do some research so that he could write and article on the matter. I contacted Mr. Zack's Foster Mother, and I subsequently passed the information on to Dr. Lant. The following article was given worldwide attention when the article was made available for publication by the Dealers of Worldprofit, of which I am one. I wrote letters to the editors of the local newspapers, but none were

From One To Infinity With Synergy

published. When I reflect on that happening, I am still bewildered by the events that took place. Although the dereliction of duty that transpired has supposedly been corrected such that it will never happen again, still there is the knowledge that the one occurrence should never have happened. I am glad that there are people who are doing whatever is necessary to remind us all of this tragic loss of life.

There is a forthcoming documentary entitled *Shallow Waters* produced by Jaime Longhi that will remind us, lest we forget.

The short life and appalling death of Raymond Zack, an avoidable American tragedy. by Dr. Jeffrey Lant Crown Memorial State Beach, Alameda, California is the kind of place you come to breathe and shake off life's trials and tribulations. The panorama is just what you think the Golden State should be...a place of possibilities, not inhibitions. Here the air is superior to any French vintage... the chill waters are bracing and playful.... Here the very birds fly higher because they are contented at such a place... and in the distance, clearly seen, is the great structure of one of mankind's signature triumphs the Golden Gate Bridge... which sends

From One To Infinity With Synergy

every spirit soaring... It was here that Raymond Zack came to die... and where the people charged with protecting life assisted Raymond take his, to the astonishment, wonder and outrage of the world. Raymond Zack, born July 23, 1959. Raymond was, like so many millions of us, a son of America's great heartland; Ohio born and bred. His life moved to the rhythm that is so quintessentially ours... He was a product of Columbus' Catholic schools... where he learned good manners, the importance of being a good man and valuable citizen... and where he glimpsed, at the hands of his dedicated instructors, the reality of God Everlasting. At 6'3" tall, this giant of a boy excelled at track and baseball... people saw him above the crowd and, with a wink and nudge, said the boy had talent. He went, and went proudly, to Ohio State.... as American as any educational establishment in the land. It was here, upon graduation, that he entered the community of educated men and women... And where he decided to answer Horace Greeley's great exhortation "Go West, young man, Go West!" And he did, attracted by the dazzling sunshine and even more dazzling possibilities of California, the pot of gold at the end of America's rainbow. But California life, for all that

From One To Infinity With Synergy

the sun was radiant, gave Raymond Zack more than his share of life's troubles. His family life was turbulent, confusing, and never restful though he was the beneficiary of his foster mother's affectionate care and unceasing concern. He weighed 300 pounds now and, like millions of his countrymen, was challenged by the complexities of food and the clear and present dangers of overindulgence. Chagrined by his bulk, Raymond, bit by bit, withdrew from the body politic and faced the secret sorrows of isolation and loneliness, the abiding reality for too many of his countrymen. His mother died in November 2010 and though there had been confusions and disappointments there, still she was his mother. Her loss magnified his burdens. Then, in the midst of a great recession, where California's profound promise was tarnished, Raymond lost his job at the St. Vincent de Paul Free Food Distribution Center where, along with Mrs. Dolores Berry, his foster mother, he had helped everyone who came.

Now the man who had helped so many was himself in need of help. This, too, was, quintessentially American for too many. Raymond, with a "God helps those who help themselves"

From One To Infinity With Synergy

attitude, tried hard to do what he'd been taught to do; to keep his chin up and a stiff upper lip; to do what he could, to stay cheerful in the face of adversity. But bit by bit, like so many, his resilience and hope were worn away. Raymond's dark days were nigh. We shall never know where Raymond's anxious forebodings carried him, alone at the midnight hour. At such a time a man may turn to booze, women, any dissipation to dispel the gloom, but Raymond seems to have faced his great matter alone and in profound despair. This, too, is reality for millions of the dispossessed and fearful. At some irrevocable moment in his profound human misery Raymond decided the game was not worth the candle and that it was time to move again, out of very life itself. Thus, on May 30, 2011, while his countrymen were celebrating the sacrifices made by others to the benefit of all, Raymond Zack decided to make a sacrifice, too: of himself, since living life was just too painful and without hope. And so he waded into the chill waters at Crown Memorial State Beach, about to be the venue of muddle, confusion, bumbling, and death. A great American tragedy was about to commence, unnecessary, scandalous, an event that enhanced no one and left Raymond Zack,

From One To Infinity With Synergy

floating face down, his life's work at an end. Seen by many. Remember, Raymond Zack was a big man, 6'3", over 300 pounds. He moved slowly, deliberately in the shallow waters. He was clearly seen though his purpose, at first, was not. Still, as Raymond walked into deeper waters, residents were concerned; a 911 call was made, alerting police and firefighters that some kind of incident was underway. In just 4 minutes help was at hand, and at hand help stayed, but without lifting a finger. And here is where an avoidable tragedy morphs into disbelief, reproach, scandal, and incomprehension. Not one of the many lifesaving professionals on the beach, not a single one, did a single thing to forestall the tragedy that could so easily have been prevented. Later these officials, pummeled by an incredulous world, worked overtime to manufacture excuses they hoped would appease, mollify and cover. Fire officials said that because of budget cuts no one knew the necessary rescue procedures. But this excuse was quickly blasted... when it was shown the department had money, but no sense. Other officials said rescue policies did not cover the case in point. A police spokesman said officers stayed out of the water because Zack was suicidal and posed a

From One To Infinity With Synergy

possible threat. A boat was requested to take officers to Zack, but those requesting it never indicated the matter was pressing. In short, at every moment where judgement, help and assistance were required, the professionals at hand, our honored paladins, were without judgement, help and assistance. And so, in full view of the world, in full view of his hysterical foster parent, 86 year old Dolores Berry, who unsuccessfully begged for celerity and assistance, Raymond Zack died.

In the way of these things, everything the system could have provided Raymond in life only emerged when he was dead; in such ways does America expiate its negligence. Now there are flowers on the beach where he died, a crowd gathers daily to reflect and wonder; bishops make Raymond the subject of their learned lamentations. Municipal officials investigate and dismiss the inept. All this is good, right and proper. But we must not forget the man at the center of it all, Raymond Zack, dead too soon at 50. He meant us well, each and every one of us. Now, prematurely, he rests in the bosom of the Lord; may he find the peace there he never had here.

From One To Infinity With Synergy

About the Author Harvard-educated Dr. Jeffrey Lant is CEO of Worldprofit, Inc., providing a wide range of online services for small and-home based businesses.

From One To Infinity With Synergy

PHOTO GALLERY

Wanda, Willis and Me
Havana, Cuba 1932

My sister Wanda

Uncle Joe Ferguson
& Family

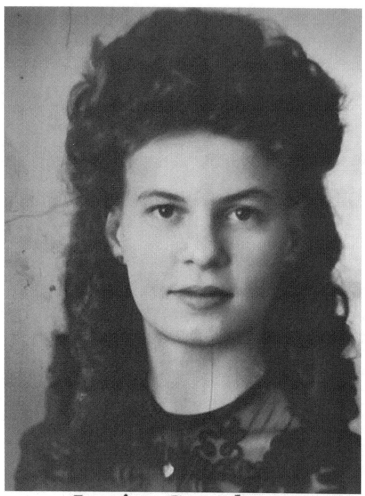

Louise Carothers
My High-school Sweetheart

Bobby & Mary Helen
1941

Wanda & Willis
1942

Mios Woendi 1943

Chief Electronics Technician
U.S.S. Hornet CVA-12

USS Jamestown AGP-3

Apollo

Gaydelle Mayon Johnson
1946

Doris & Wallace
1954

Doris and Wallace
40th Anniversary 1994

Joan Lindsey
2008

Wallace A. Johnson MBA
President & CEO
Dewaj Synergy International

Wallace A. Johnson MBA
www.IHaveLiftOff.com

From One To Infinity With Synergy

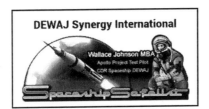

From One To Infinity With Synergy